YOUR KNOWLEDGE HAS VALUE

- We will publish your bachelor's and
  master's thesis, essays and papers

- Your own eBook and book -
  sold worldwide in all relevant shops

- Earn money with each sale

Upload your text at www.GRIN.com
and publish for free

Björn Linnemann

Murray-Darling Basin: Human impact on the ecosystem and consequences for water resource management

GRIN Verlag

Bibliografische Information der Deutschen Nationalbibliothek:

Die Deutsche Bibliothek verzeichnet diese Publikation in der Deutschen National-
bibliografie; detaillierte bibliografische Daten sind im Internet über http://dnb.d-
nb.de/ abrufbar.

Dieses Werk sowie alle darin enthaltenen einzelnen Beiträge und Abbildungen
sind urheberrechtlich geschützt. Jede Verwertung, die nicht ausdrücklich vom
Urheberrechtsschutz zugelassen ist, bedarf der vorherigen Zustimmung des Verla-
ges. Das gilt insbesondere für Vervielfältigungen, Bearbeitungen, Übersetzungen,
Mikroverfilmungen, Auswertungen durch Datenbanken und für die Einspeicherung
und Verarbeitung in elektronische Systeme. Alle Rechte, auch die des auszugsweisen
Nachdrucks, der fotomechanischen Wiedergabe (einschließlich Mikrokopie) sowie
der Auswertung durch Datenbanken oder ähnliche Einrichtungen, vorbehalten.

Imprint:

Copyright © 2010 GRIN Verlag GmbH
Druck und Bindung: Books on Demand GmbH, Norderstedt Germany
ISBN: 978-3-656-25227-6

This book at GRIN:

http://www.grin.com/en/e-book/198580/murray-darling-basin-human-impact-on-
the-ecosystem-and-consequences-for

GRIN - Your knowledge has value

Der GRIN Verlag publiziert seit 1998 wissenschaftliche Arbeiten von Studenten, Hochschullehrern und anderen Akademikern als eBook und gedrucktes Buch. Die Verlagswebsite www.grin.com ist die ideale Plattform zur Veröffentlichung von Hausarbeiten, Abschlussarbeiten, wissenschaftlichen Aufsätzen, Dissertationen und Fachbüchern.

Visit us on the internet:

http://www.grin.com/

http://www.facebook.com/grincom

http://www.twitter.com/grin_com

<u>Murray-Darling Basin:</u>
<u>Human impact on the ecosystem and</u>
<u>consequences for water resource</u>
<u>management</u>

Hausarbeit

Vorgelegt von:

Björn Linnemann

Institut für Geographie - Universität Hamburg
M Sc Geographie

CONTENTS

<u>LIST OF FIGURES</u>

1 Scope and aim of the paper

The Murray Darling Basin, located in the south eastern part of Australia, is the countries largest catchement area and main water reservoire for the cities and the agriculture. Historically and now even more in the course of climate change, Australia is facing significant water shortages. The Murray Darling Basin is running out of water and thus challenging the habits of Australians in the cities and in rural agriculture. The tense situation is a challenge in environmental, socio-economic and cultural regards. Understanding the problematical situation and its historical development over time as well as developing measures to be taken in order to improve conditions in a sustainable way is demanding a wide scope and persepective. Revealing and describing the different factors and the complex interdependencies requires an integrative approach.

The aim of this paper is to describe the specific situation of the Murray Darling Basin and to put it into an integrative context. This integrative context will deliberately embrace several concepts to show the multiplicity of the situation. In this specific case I examine the concept of human ecology for a wide understanding of the problem of water shortage itself. In the further discussion about current problems and challenges of water resource management I will take a closer look at the concept of environmental justice. Based on that, I will figure out the weaknesses of current measures taken by the stakeholders in Australia.

2 Australia's climate situation

Australia is a large continent spanning tropical, sub-tropical and temperate climatic zones. Temperature trends in Australia over the past century are consistent with global trends in showing a more or less steady warming. Australia is the driest inhabited continent on earth and its water resource issues are intimately related to prevailing rainfall distribution. The availability of water is thus more critical than a rise in temperature per se and crucial to maintain the pattern of habitation and agriculture (Steffen et al., 2010; Risbey, 2010).

The northwestern part of the country has become wetter while a pronounced drying trend has appeared in the south west and along the east coast, the region where most of the Australian population lives and where much of the agricultural activity occurs (see Figure 1, Steffen et al., 2010). Furthermore eastern Australia is subject to effects of the El Niño-Southern Oscillation phenomenon, experiencing major droughts during these periods (Steffen et al, 2010).

An additional concern is that the southern coast and cities (Perth, Adelaide, Melbourne, Hobart, Sydney) are dependent on the passage of frontal systems to provide much of their rainfall. If these systems were to contract to the south as a result of greenhouse or ozone forcing of the climate (Hartmann et al. 2003; Karoly 2003), there is potential for major rainfall reductions in these regions.

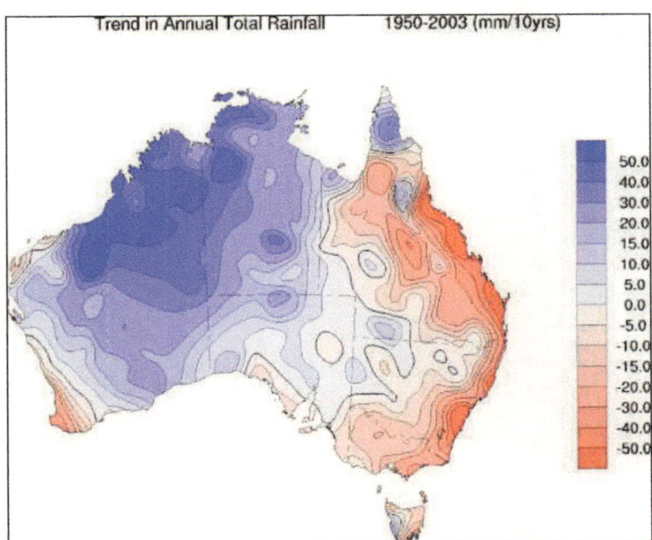

Figure 1: Trend in mean annual rainfall
Source: Australian Bureau of Meterology/Risbey, 2010

Though rainfall is hard to project in models, one of the consistent results that emerges is that the dry regions tend to get dryer and wet regions wetter under greenhouse scenarios (Hansen et al. 1998; Houghton et al., 2001). With the second highest rate of CO_2 emissions per capita and 4.5 times the global average Australia is further contributing to climate change and thus boosting the risk of growing water shortage (Preston and Jones, 2006). Consequently the water supply situation is likely to worsen in the future and resources may be the 'Achilles heel' of Australia under climate change (Risbey, 2010).

3   The Murry Darling Basin

The Murray Darling Basin is by far the most significant agricultural area in Australia. The name is derived from the two largest rivers Murray River and Darling River. The largest part of the basin is located in the state of New South Wales, spanning into Queensland, Victoria, South Australia and the considerably small Autralian Capital Territory (see Figure 2).

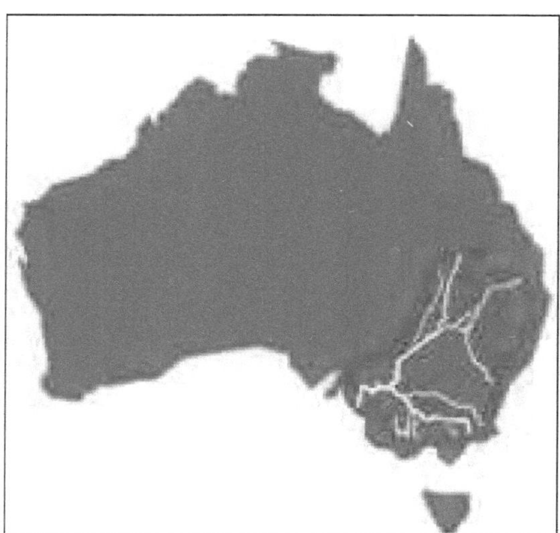

Figure 2: Location of the Murray Darling Basin
source: Risbey, 2010

The basin is the largest catchment area of Australia and the main source of water supply for most of the country's major cities and the lion's share of the nation's agriculture.

Agricultural production follows the rainfall regions and the major river valleys that spring from these regions. Major agriculture thus favours the east and south coasts and inland regions such as the Murray Darling basin (Risbey, 2010).

Irrigation in Australia is a widespread practice to supplement low rainfall levels in Australia with water from other sources. Most of Australias cropping and irrigated agriculture is located in and around the Murray Darling Basin (see Figure 3).

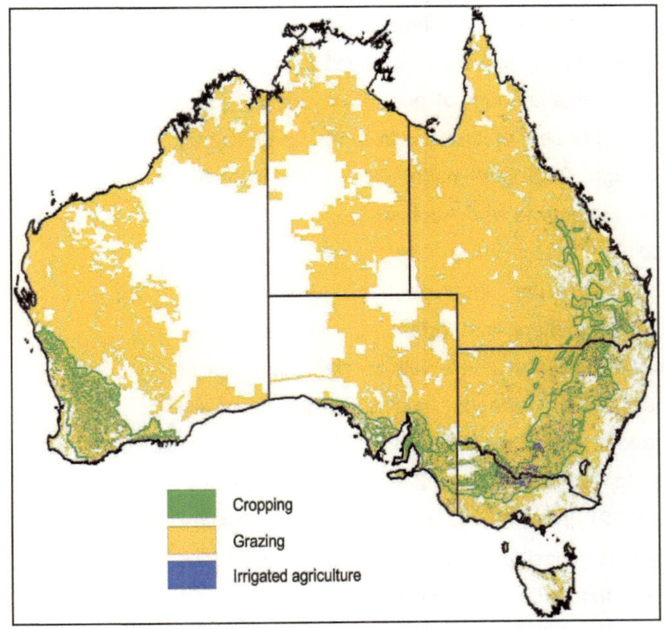

Figure 3: Map of land use across Australia and major agricultural zones
source: Steffen et al, 2010

Characterised by its vast flatness, rain falls into large storages located along the eastern mountainous border, and drains south west through multiple interconnecting networks of natural water courses and into man-made irrigation canals. The Murray Darling Basin accounts for roughly 34 per cent of Australia's agricultural production by gross national value and 75 per cent of its irrigated area. Of the approximately 13,000 gigalitres of flow in the basin, which studies have shown to be divertible, 11,500 gigalitres per annum is removed for irrigation, industrial use, and domestic supply (Bentley, 2007).

Agriculture is a major backbone of the Austrailan economy.and droughts cause significant effect on GDP. The 2002-2003 drought for example resulted in an estimated drop of 1,6 per cent in GDP loss of 70.000 jobs (Steffen et al., 2006).

Besides irrigated agriculture the major cities like Sydney, Melbourne, Brisbane and Adelaide do receive a large amount of their drinking water. The latter for example gets a normal annual rainfall of 545.3 mm (BoM, 2009). For comparison the German city of Munich receives approximately 849 mm per year for a comparable amount of population. In a drought year Adelaide sources up to 90 per cent of its water from the River Murray (Department of Climate Change, 2009). All major cities have instituted major water restrictions, with Melbourne and Adelaide declaring some of them permanent (Risbey, 2010).

The decrease in rainfall over the last decades, the massive overexploitation of water for agriculture and urban use plus effects of climate change resulted in sinking water levels with partial drying out of the rivers and thus significant water shortages. Inflows into the basin are currently the lowest on record and and river flows in the basin are expected to decrease by 16 to 48 per cent for a 3-4°C warming (Bentley, 2007; Beare and Heaney, 2002). If flows in Murray Darling river system halve, the agricultural mix is considered not be sustainable. The Commonwealth Scientific and Industrial Research Organisation (CSIRO) has forecasted that climate change will cause decreased precipitation over much of Australia and that this will exacerbate existing challenges to water availability and quality for agriculture (Preston and Jones, 2006).

Overuse and poor irrigation practices have furthermore led to increased salt content in the soil, reducing the productivity of the soil.

The loss of environmental flows in river systems often reduces water quality and exacerbates salinity stress on riverine ecosystems. Climate change as it is forecasted would make these kinds of impacts more frequent and severe if not a normal case (Chiew and McMahon, 2001).

The remaining water in the waterways is often contaminated with the wash out of fertilizers such as phosphorus and nitrogen (Australian State Of Environment Committee, 2006). The growing shortages of water supply intensifies the already existing competition for the resources of the river systems in the Murray Darling Basin between Queensland, New South Wales, Victoria, and South Australia.

4 Measures taken for mitigation of water shortage

It is broadly agreed that water in Australia has been over-allocated resulting in social and environmental degradation (Cullen, 2002). Since European settlers arrived in Australia in the late 1700s, the availability of water for towns, agriculture and later industry has been a continuing and predominant concern and noumerous initiatives to adequately manage the water resources have been undertaken since (Gross, 2008). In the early days of European settlement, water resources were not controlled. To gain control over the water allocation and the regulation of the Murray River system was one of the first issues addressed after Federation in 1901. After intensive negotiations the River Murray Waters Agreement was signed in 1915. The agreement to which the Commonwealth and the states of New South Wales, Victoria and South Australia were parties, set out the basic conditions, which remain in force today. It provided for construction of dams, weirs and locks on the main stream of the Murray to be managed by the River Murray Commission, which was established in 1917. Escalating problems and challenges over time lead to new negotiations in the 1980s resulting in the Murray Darling Basin Agreement replacing the River Murray Agreement. The new agreement was putting in place a process for the effective management of the water, land and other environmental resources on a basin-wide basis (Eastburn, 1990). Its purpose is to promote and coordinate effective planning and management for the equitable, efficient and sustainable use of the water, land and other environmental resources of the Murray–Darling Basin. In parallel the Murray-Darling Basin Commission was founded replacing the old River Murray Commission managing the country's water resources.

Initially the authorities sought to gain control over water allocation and consumption in agrilculture, but quickly it became obvious that water management would have to be applied for the rapidly growing cities. This lead to water cuts being implemented in the cities especially during dry periods. Abuse or misuse of water is now punishable by law.

Effectively the development led to all rights to use and control water being vested in the state nowadays. Users are then issued various conditional entitlements to use water and some of these entitlements can, in limited circumstances, be traded.

Numerous other institutions and initiatives were put in place to manage to shortage of water supply in the country. The National Water Commission for example is responsible for driving progress towards the sustainable management and use of Australia's water resources and makes recommendations for actions. The so-called Murray-Darling Cap is a policy implemented in 1997, limiting the water diversions in the Miurray Darling Basin at 1993 levels. It seeks to strike a balance between the amount of water available to irrigators and the security of their water supply. At the time of conception the level of diversions from the basin was clearly nearing unsustainable levels. The Cap effects all rivers located within the Murray-Darling Basin. The following years saw increased trade in these entitlements.

Prior to the formal establishment of markets in the mid-1990s, water trading was often an informal neighbourly exchange. With the establishment of formal rights over the past 10 to 15 years, the market has evolved rapidly. There are several forms of trade of water entitlements. Firstly, the permanent water entitlement can be traded – known as permanent trade. Secondly, once a water announcement is made, the annual yield on a permanent entitlement can be traded, known as annual or temporary trade. Annual trading is by far the more active market and, in some irrigation districts, has equated to up to 30 per cent of all water used on farms (Bentley, 2007).

With the potential for the drought to worsen, there is growing pressure upon government to look towards non-market mechanisms to allocate water (Bentley, 2007). In 2004 National Water Initiative (NWI) set out a comprehensive national water reform process, which aims to rework the current management (Gross, 2008).

What all applied measures until now have in common is, that all the restrictions put in place by central institutions are trying to meet water shortage with installation of technological means and cuts of water entitlements.

5   The Murray Darling Basin from an integrated perspective

With the specific interplay of nature and mankind with a massive overuse of water in combination with dangers of climate change, the case of the Murray Darling Basin serves as a depictive piece of environmental change caused by both mankind and natural phenomena. Hence, an integrative approach in order to understand the emergence of certain events (like massive droughts in the basin) and to derive adequate actions or initiatives seems to be required.

It is not easy to draw an accurate line in the current discussion about the interplay between nature and mankind. The boundaries between different concepts are often blurry. Taking this into account I will start examining the situation from the wide angle of human ecology shaping and narrowing the scope while walking the integrative path.

Human ecology is dedicated to the research of cause and effect connections and interactions between, society, man and nature (Weichhardt, 2007). It is furthermore an action-based science, which monitors and reflects the human actions in the scope of the natural surroundings (Glaser, 1989). Human actions in the frame of climate challenges are the center of attention in our case. The Murray Darling Basin has been under human influence since the arrival of Aboriginals

50.000 years ago. During that time, there is good evidence that they have been associated with rivers and lakes and exploited a vast range of their resources (Bandler, 1995). As a major source for water, it has since been a central point of where humans interact directly with the ecosystem.

Human impact or influence in the ecosystem is often subtle though. Many actors contributing little factors can sum up to a major effect. It is mostly difficult to identify the specific human impact against a background of natural variation (Turner and Meyer, 1993). This is especially the case for climate change the world is experiencing now. The global warming is considered to be induced by man-made CO_2 emissions, but cannot be traced back to one single factor. Man-made and natural effects cannot to be distinguished cleary. The human system influences the ecosystem and vice versa. There are actions and reactions on both sides. Subtle human imapcts are easily hidden in a broad spectrum of continious and erratic developement and changes (Strayer et al., 1986).

With the economic utilization of its rich resources of carbon Australia unquestionably contributes significantly to CO_2 emissions and therewith to global warming (CSIRO, 2007). As for the global warming is considered a major driving force in the nations drying trend and growing water scarcity, Australia not only has to deal with climate change, but is shaping it to a certain extend. Thus the climate trends themselves have to be considered a natural and human phenomenon.

Especially in the case of the Murray Darling Basin there are multidimensional reasons and effects of emerging natural phenomena. The rivers tend to dry out as a reaction to overuse and climate change. Humans then try to adapt to the circumstances and cope with them. They react on the rivers drying out. Nature again changes under the influence of human measures for adaption. Different techniques for irrigation or water use for example would change the detailed design of the ecosystem in a different way. We can refer to this as an action-reaction-chain. There are interdependancies within the ecosystem itself (e.g. waterways and precipitation or hydrospheric influence on the climate an vice versa) and links between humans and the ecosystem. This is not to say that there is a distinguished human system and an ecosystem. The waterways, precipitation, humans and multiple other factors rather form one complex system (Grossmann, 1993).

Grossmann (1993) defines three layers of a system. Firstly, there is the lowest layer; the layer of details. This layer consists of details that comprise a system. In our case, we can distinguish between the above mentioned factors mankind, waterways and precipitation for example being components of a system. Secondly, there is the middle layer of complex dynamics and interrelationship. This layer describes the dynamics imposed by lowest layer. This means that the interplay of the system components generates certain dynamic processes. In the case of the Murray Darling Basin we would refer to the interplay between human agricultural activity, natural resource provisions, human greenhouse gas contribution, wheather phenomena and so on. Thirdly, on top there is the strategic layer that mirrors the overall development of a complex system such as the Murray Darling Basin area or the Australian nation as a whole dealing with drying trends and nationwide water shortage in all respects (Grossmann, 1993). As for the complex system as a whole is always mirrored in the topmost layer, this layer would consequently picture a landscape. This landscape would consist of the natural landscape and the cultural system (including the industrial system as an emergence of human culture) merging to one cultural landscape. The cultural system of a nation is dependant on its natural landscape (Glaser, 1989). One can say that cultural and natural factors are reinforcing each other into one cultural landsape.

The Australian nature with its natural wonders and rough terrains is iconic to the nation and its inhabitants. It plays a crucial role in the nation defining itself. The conditions a nation is founded upon are formative for the self-conception of a nation and consequently its habits and attitude towards nature. Early European settlers found Australia to be a rough and dry continent with an abundance of natural resources to be exploited under extremely harsh conditions. They had to tame the land before they were able to live off it. Despite the nature's iconic status, the perception of gaining control of the rough environment seems to prevail until now. Tackeling the natural challenges with man-made technology nowadays is something of a cultural heritage in order to maintain control over nature. Technical means, such as dams or weirs are used to generate energy or control water flows. The often-negative impact of mankind on the natural landscape in Australia is not limited to the European settlers. Shortly after the Aboriginals settled on the continent, several mammals were extinct (Nentwig, 2005). Nevertheless the arrival of European settlers marked the starting point, when they replaced the human role of a hunter in the ecosystem played by the Aboriginals with a system mass consumpton and therefore massive use of resources. Ecosystems with human participation evolve with settlement and high populations (Boyden, 1993). Environmental change has escalated since human numbers have grown and other aspects of human society have changed like agriculture and industrialisation. The resulting social organisation and technological progress lead to changes in the ecosystem (Turner and Meyer, 1993).

According to Nentwig (2005), Australia's size of agricultural area grew until the 1970s and then decreased until 2000, whereas irrigation increased dramatically. Approximately 50 to 60 per cent of the growth in agricultural production between 1960 and 1980 can be traced back irrigation (Nentwig, 2005). The country's natural assets are the very backbone to the national economy, making especially water a matter of specific national interest.

Keeping all these findings in mind, we see that a massive number of stakeholders in this highly indutrialized country are involved. Environmental issues would not just come up, if it were not for the interests of certain groups triggering social processes (Blaikie, 1999). This very fact marks the need for political ecology, if it is not even intrinsic to such an accumulation of interests. Within this frame the question of environmental justice emerges: Is the principle of water cuts considered as righteous by all stakeholders? By dealing with the weaknesses of the current water management in more detail, I want to shed a light on how some of the involved stakholders perceive justice of the methods and processes in a specific case.

6 Weaknesses of current water management

The permanent concern about water shortage has stimulated a multifaceted and complicated debate in the Australian public raising critical questions towards the government's approach in water management (Gross, 2008).

As alredy stated in section 4 Australian water management seems to concentrate central authorities trying to meet water shortage with installation of technological means and cuts of water entitlements. The search for alternative measures and technologies seems to largely neglect interations between society, technology and ecology.

In a research study following the 2006 drought Gross (2008) focused on a major problem of water management in Australia.

The latent tension concerning the allocation of water ressources broke the surface on December that year in the small rural town of Deniliquin in New South Wales. The community is fed by irrigated water from waterways of the Murray Darling Basin. Over 2000 members of the local community took to the streets protesting against recent government actions. In October decision-making authorities in New South Wales unexpectedly cut part of Deniliquin's seasonal water allocation without offering compensation during an extended period of drought. The first cutback reduced the allocation by 20 per cent and the second cutback by a further 32 per cent. So eventually local farmers ended up with only 48 per cent of the assigned water entitlements. Farmers were protesting about a perceived property right being taken away from them without adequate recognition or compensation (Gross, 2008). Interviews Gross' conducted in the community after the incident revealed a sense of unfairness in the way the irrigation community had been treated. The reaction of the people of Deniliquin and the fact, that water is considered a national treasure and public good, the factor of public acceptance of government's restrictions, laws and water cuts seems to have been underestimated by the authorities. The water cuts were judged to be unfair and wrong by most people of the community. Nevertheless, some people expressed different perspectives about water allocation and the irrigation community reaction.

Depending on the several roles people had in the community certain interests dominated their position (Gross, 2008). Whether something is seen to be fair or unfair depends on which of the roles is foremost in a certain case (Skitka, 2003). Syme and Nancarrow (2005) argue that justice and fairness are important components of decision-making processes that can result in greater acceptance of outcomes. One accepts what he considers fair and just. Perceived justice is formed from the consideration of distribution, recognition and procedure (Schlosberg, 2003). Within the factors of procedure and recognition the aspect of participation is crucial for public acceptance. 'Am I recognised as a stakeholder and am I actively involved in the process?' is one of the central questions to effected individuals or groups. Justice includes the ability to participate and access to adequate information (Tyler, 2000).

Most certainly this participation and thus justice and fairness were not given in the case of the forcefully imposed water cuts in Deniliquin. As for water cuts are a common procedure by authorities during drought, one can assume a general lack of perceived justice and fairness within parts of Australian water management. Remembering the layers defined by Grossmann (1993), current water management seems not to have wholly comprised the layer of details and within this values and needs of the population.

To gain general acceptance means that a topic has to be researched from different perspectives. Authorities cannot simply refer to the short water resources in order to take action; the cultural landscape as a whole needs to be considered. Furthermore, different stakeholders within the cultural landscape need to be integrated into the decision-making process or in the forming of a water management system.

The participation of people in the procedures might as well be describes as a kind of 'environmental citizenship', a phrase formed by Dobson (2007). This citizenship would carry a duty- or responsibility-oriented component along with rights. This takes the factors of distribution, recognition and procedure one step further. Stakeholders do not only have certain rights, they are also bound to certain duties. This is central to a widely accepted and perceived responsibility among all involved parties.

The causal chain of responsibility and rights inherently carries reinforcing elements. Having certain rights makes people actively accept responsibility. Having responsibilities makes people ask for rights in return. Consequently a sensitisation for the own acting will be achieved.

7   Conclusion

The Murray Darling Basin is a vivid example of human action as part of an ecosystem. The existing problems and challenges have to be observed in a wider frame, taking into account the complex ecosystem with all its members and multiple interrelationships as a whole.

This paper has investigated the current methods of Australian water management and the actions of the aurthorities in more detail taking into account the historical context. It has considered the climate situation and prospects in the frame of human ecology. We also took into account the position and meaning of nature within the Australian culture and the evolution of water management and policy over time.

The challenges for a water policy that finds broad acceptance and civil cooperation can be derived more or less directly from the main weaknesses of current water management. The paper discussed these weaknesses on the basis of perceived environmental justice with special attention to participation.

Farmers for example work in an inseparably unit consisting of social, economical and physical environment. Their work is shaped by the physical conditions like soil or precipitation, the needs of their families, the market demand for certain agricultural products and its price, the interaction with the authorities and other farmers (Kölsch, 1989). These factors among a wider range preconditions including other stakeholders and their roles have seemingly been largely neglected by governmental policy and consequently its actions lacked acceptance among the population.

In order to gain broad acceptance, the authorities have to respect the different members of the complex cultural landscape and closely integrate them in the water management process. This also includes the indengenious people of Australia. The natural expertise of Aboriginals is valued for aboriginal co-management of National Parks and National Reserves since recent decades. This is a broad practice throughout all states. Aboriginal involvement in managing water resources is still little and needs to be applied at a larger scale. Their often deep traditional knowledge about landuse would be useful.

A high degree of transparency and integration of the aggrieved parties has another positive effect. By consulting and integrating them, authorities have access to a wide spectrum of specific local knowledge. This knowledge provides them with a firmer basis for decisions and actions in return.

Australian farmers on average, are well-placed to respond to climate change, having access to R&D innovation, agribusiness services, education services, modern infrastructure, a range of marketing and storage systems, and are well-served by financial markets (Steffen et al., 2006). Furthermore farmers and rural population are becoming more aware of climate change and more interested in adapting to its potential impacts (Steffen et al., 2010). The willingness to get involved in mitigating the threats of water shortage among is thus given. This is a major advantage for democratizing the water management and having it backed and supported by the broad population. This is actually close to a kind of environmental citizenship and a positive perception of environmental justice.

REFERENCES

Australian State Of Environment Committee (2006): The State of environment Report 2006, URL: http://www.environment.gov.au/soe/2006/publications/report/pubs/soe-2006-report.pdf, retrieved: 2010-01-04.

Bandler, H. (1995): Water resources exploitation in Australian prehistory environment, in: The Environmentalist 15, 97–107.

Beare, S. and Heaney, A. (2002): Climate change and water resources in the Murray Darling Basin, Australia: impacts and adaptation, ABARE Conference Paper 02.11.

Bentley, J. (2007): Testing the limits of trade, in: Environmental Finance, July-August 2007, 32-35.

Blaikie, P. (1999): A review of political ecology - Issues, epistemology and analytical narratives, in: Zeitschrift für Wirtschaftgeographie 43, 131-147.

BoM (Australian Bureau of Meteorology) (2009): Living with drought, URL: http://www.bom.gov.au/climate/drought/livedrought.shtml, retrieved: 2009-12-12.

Chiew, F. and McMahon, T. (2001): Modelling the impacts of climate change on Australian streamflow, in: Hydrological Processes, 16, 1235–1245.

CSIRO (Commonwealth Scientific and Industrial Research Organisation) (2007): CO2 emissions increasing faster than expected, URL: http://www.csiro.au/news/GlobalCarbonProject-PNAS.html, retireved: 2010-01-04

Cullen, P. (2002): The common good, in: Connell, D. (editor): Uncharted Waters, Murray-Darling Basin Commission, 49-60.

Deparmtent of Climate Change, Australian Government (2009): Murray Darling Basin, URL: http://www.climatechange.gov.au/climate-change/impacts/national-impacts/murray-darling-basin.aspx, retrieved: 2009-12-12.

Dobson, A. (2007): Environmental Citizenship: Towards Sustainable Development, in: Sustainable Development 15, 276–285.

Eastburn, D. (1990): The River Murray History at a Glance, Murray-Darling Basin Commission.

Glaser, B. (1989): Entwurfe einer Humanökologie, in: Glaser, B. (editor): Humanökologie: Grundlagen präventiver Umweltpoitik, Opladen, 27-45.

Gross, C. (2008): A Measure of Fairness: An Investigative Framework to Explore Perceptions of Fairness and Justice in a Real-Life Social Conflict, in: Human Ecology Review, Vol. 15, No. 2/2008, 130-140.

Grossmann, W. D. (1993): Integration of Social and Ecological Factors: Dynamic Area Model of Subtel Human Influences on Ecosystems Integration, in: McDonnell, Mark J., Pickett, Stewart T.A. (editors): Humans as Components of Ecosystems – The Ecology of Subtle Human Effects and Populated Areas, New York, 229-245.

Hansen, J., Fung, I., Lacis, A., Rind, D., Lebedeff, S., Ruedy, R., Russell, G., and Stone, P. (1998): , in: Journal for Geophysical Research, 93, 9341–9364.

Hartmann, D., Wallace, J., Limpasuvan, V., Thompson, D., and Holton, J. (2003): Can ozone depletion and global warming interact to produce rapid climate change, in: Proc. Nat. Acad. Sci. USA, 97(4), 1412–1417.

Houghton, J. T., Ding, Y., Griggs, D. J., Noguer, M., van der Linden, P. J., Dai, X., Maskell, K., and Johnson, C. A. (2001): Climate Change 2001: The Scientific Basis, Cambridge, 881.

Karoly, D. (2003). Ozone and climate change, in: Science, 302, 236–237.

Kingwell, R. (2006). Climate change in Australia: agricultural impacts and adaptation. Australasian Agribusiness Review 14: 1.

Kölsch, O. (1989): Humanökologische Forschung für Landwirtschaft und Agrarpolitik, in: Glaser, B. (editor): Humanökologie: Grundlagen präventiver Umweltpoitik, Opladen, 181-193.

Nentwig, W. (2005): Humanökologie: Fakten-Argumente-Ausblicke, Bern.

Preston, B.L. and Jones, R.N. (2006): Climate Change Impacts on Australia and the Benefits of Early Action to Reduce Global Greenhouse Gas Emissions, CSIRO Marine and Atmospheric Research, URL: http://www.csiro.au/files/files/p6fy.pdf, retrieved: 2010-01-04.

Risbey, James (2010): Dangerous climate change and water resources in Australia, in: in: Hare, W. L., Battaglini, A., Cramer, W., Schaeffer, M. and Jaeger, C. (editors): Climate hotspots: Key vulnerable regions, climate change and limits to warming, Heidelberg, 91-98.

Schlosberg, D. (2003): The justice of environmental justice: Reconciling equity, recognition and participation in a political movement, in: Schlosbere, A. and de-Shalit, A. (editors): Moral and Political Reasoning in Environmental practice, London, 77-106.

Skitka, L.J. (2003): Of different minds: An accessible identity model of justice reasoning, in: Personality and Social Psychology Review 7, 4, 286-297.

Steffen, W. S., Sims, J., Walcott, J. and Laughlin, G. (2010): Key Vulnerable Regions and Climate Change: Australian Agriculture; in: W. L. Hare, A. Battaglini, W. Cramer, M. Schaeffer, C. Jaeger (editors): Climate hotspots: Key vulnerable regions, climate change and limits to warming, Heidelberg, 112-131.

Steffen, W., Love, G. and Whetton and P. (2006): Approaches to defining dangerous climate change: a southern hemisphere perspective, in: Schellnhuber, H.J., Cramer, W., Nakicenovic, N., Wigley, T. and Yohe, G. (editors): Avoiding Dangerous Climate Change, Cambridge, 219-225.

Strayer D., Glitzenstein J.S., Jones C.G., Kolasa J., Likens G.E., McDonnell, M.J., Parer, C.D. and Pickett, S.T.A. (1986): Long term ecological studiess: an illustrated account of their design, operation and importance to ecology, in: Occ. Publ. Inst. Ecosys. Stud., 2, 1-38.

Syme, G. and Nancarrow, B. (2005): Creating community consent through fairness judgements, in: D. Cryle and J. Hillier (editors.): Consent and Consensus: Politics, Media and Governance in Twentieth Century Australia, Pert, 371-387.

Turner, B.L. II and Meyer, W. B. (1993): Environmental Change: The Human Factor, in: McDonnell, Mark J., Pickett, Stewart T.A. (editors): Humans as Components of Ecosystems – The Ecology of Subtle Human Effects and Populated Areas, New York, 40-60.

Tyler, R.T. (2000): Social justice: Outcome and procedure, in: International Journal of Psychology 35, 2, 117-125.

Weichhardt, P. (2007): Humanökologie, in: Gebhardt, H., Glaser, R., Radtke, U., Reuber, P. (editors): Geographie: Physische Geographie und Humangeographie, München, 941-949.